Key to
Algebra®

ADDING AND SUBTRACTING RATIONAL EXPRESSIONS

7
Student Workbook

By Julie King and Peter Rasmussen

Name _____ Class _____

TABLE OF CONTENTS

Young and Gifted II

Many mathematicians tried to solve the general fifth degree equation from 1545 until 1820, when the teenager Niels Henrik Abel succeeded where they had failed. He showed that it is impossible to construct a formula to solve every equation of degree five or higher.

Meanwhile in France another child prodigy met a similar fate.

Evariste Galois (pronounced *gal-WAH;* 1811-1832) led a rebellious life. He didn't attend school until he was 12 years old; his mother provided him with a classical education until then. Evariste was only an average student until he took a math course whose textbook by the famous mathematician Legendre caught his fancy. In two days he had finished the book designed for two years of study.

Galois then handed in his textbook with the explanation, "I know it." He was right. When Galois said he knew something, he meant he had mastered it.

His teacher, Louis-Paul-Emile Richard, played an important role in his life, particularly since Galois didn't get along with many other students at the boarding school. Mr. Richard arranged for Galois to take the examination for entrance into France's most famous university, the Polytechnic Institute. Unfortunately the young Galois failed the oral part of the exam when he refused to answer a question he felt was improperly posed.

Do you think that Galois' attitude was impertinent? Should he have challenged the experts? Most mathematicians today agree that his opinion was correct, even if his behavior wasn't right.

So Galois returned to high school and worked independently under Mr. Richard. Soon, however, he became involved in radical politics and at one point made a threat on the king's life. He was sent to prison for the threat, and he remained there until ill health forced the authorities to transfer him to a sanitorium.

While there he was challenged to a duel. He accepted.

That was very stupid. Galois had no experience with pistols. The next morning he was killed in the duel. He was only 20 years old.

Historical note by
David Zitarelli

Illustration by
Jay Flom

That is very sad.

Fortunately Galois stayed up the entire night before the duel writing a letter to his best friend. In it he summarized his results in algebra and outlined the direction that his research would lead.

Without knowing of Abel's work Galois proved exactly the same result—it is impossible to construct a formula to solve fifth degree equations. But he went even further. He knew that some equations could be solved and others could not. He introduced groups of objects to determine whether a given equation of any degree could be solved or not.

Galois was ahead of his time. His works were not understood for some 25 years after his death. Today graduate courses in "abstract algebra" begin by studying group theory and end by studying "Galois theory."

On the cover of this book Galois prepares for his duel. Unlike the mathematical "duels" of the Renaissance, this duel was fought with pistols and brought the end to Galois' short life.

Copyright © 1990 by Key Curriculum Project, Inc. All rights reserved.
® *Key to Fractions, Key to Decimals, Key to Percents, Key to Algebra, Key to Geometry, Key to Measurement,* and *Key to Metric Measurement* are registered trademarks of Key Curriculum Press.
Published by Key Curriculum Press, 1150 65th Street, Emeryville, CA 94608
Printed in the United States of America   23   08   ISBN 978-1-55953-007-1

Review

In this book you will learn how to add and subtract fractions (both rational numbers and rational expressions). To do this you will have to remember how to add, subtract, multiply and factor polynomials, how to simplify rational expressions and how to find equivalent expressions in higher terms. Test your memory by doing these problems.

Simplify.

$$\frac{3a}{3b} =$$

$$\frac{3a + 3b}{a + b} =$$

$$\frac{x-4}{x^2-16} =$$

$$\frac{2x+8}{x^2+7x+12} =$$

Find an equivalent fraction.

$$\frac{x}{y} = \frac{}{5y}$$

$$\frac{x}{12} = \frac{}{12xy}$$

$$\frac{3}{x+2} = \frac{}{(x+2)(x-1)}$$

Add.

$$(x^2 + 4x - 6) + (2x^2 + x - 8) =$$

$$(x^2 + 5) + (x - 1) =$$

Write the opposite of the expression in parentheses.

$$-(x^2 + 8) =$$

$$-(3x^2 - 5x + 2) =$$

Subtract.

$$3x - {}^-5x =$$

$$(3x^2 + 5x) - (x^2 + 4x) =$$

$$^-2a^2 - 6a^2 =$$

$$(x^2 + 7x + 8) - (x^2 - 2x - 5) =$$

Multiply. Make sure your answer is in simplest form.

$$(x + 7)(x - 1) =$$

$$2x^2(x - 5) =$$

$$8x^2 \cdot \frac{1}{2x} =$$

$$5y^2\left(\frac{1}{y} + \frac{2}{5}\right) =$$

Factor completely.

$$105 =$$

$$3x^2 - 27 =$$

$$168 =$$

$$x^2 + 3x - 88 =$$

Adding Fractions with a Common Denominator

When two fractions have the same denominator we say that they have a **common denominator**. Adding fractions with a common denominator is easy. Denominators tell *what kinds* of numbers or expressions are being added. Numerators tell *how many,* so to add fractions we just add the numerators and keep the denominators the same.

$$\frac{3}{4} + \frac{2}{4} = \frac{5}{4} \qquad\qquad \frac{4}{a} + \frac{-7}{a} = \frac{-3}{a}$$

Add.

$\frac{5}{8} + \frac{2}{8} =$	$\frac{1}{5} + \frac{3}{5} =$	$\frac{2}{9} + \frac{3}{9} =$
$\frac{-7}{6} + \frac{-4}{6} =$	$\frac{16}{3} + \frac{-2}{3} =$	$\frac{1}{8} + \frac{1}{8} + \frac{3}{8} =$
$\frac{1}{x} + \frac{4}{x} =$	$\frac{5}{a} + \frac{3}{a} =$	$\frac{1}{5b} + \frac{3}{5b} =$
$\frac{4}{7y} + \frac{5}{7y} =$	$\frac{-6}{x^2} + \frac{1}{x^2} =$	$\frac{7}{xy} + \frac{-4}{xy} =$
$\frac{1}{x+3} + \frac{4}{x+3} =$	$\frac{12}{a-1} + \frac{3}{a-1} =$	$\frac{-7}{x^2-8} + \frac{2}{x^2-8} =$
$\frac{n}{3} + \frac{5}{3} =$	$\frac{x}{4} + \frac{y}{4} =$	$\frac{a}{5} + \frac{-3a}{5} =$
$\frac{4x}{3} + \frac{3y}{3} =$	$\frac{a^2}{9} + \frac{b^2}{9} =$	$\frac{-x}{3} + \frac{8x}{3} =$

2

Sometimes after adding we can simplify the answer.

$$\frac{8}{9} + \frac{7}{9} = \frac{\overset{5}{\cancel{15}}}{\underset{3}{\cancel{9}}} = \frac{5}{3}$$

$$\frac{-2x}{5y} + \frac{7x}{5y} = \frac{\overset{1}{\cancel{5}}x}{\underset{1}{\cancel{5}}y} = \frac{x}{y}$$

$$\frac{x}{x^2-4} + \frac{2}{x^2-4} = \frac{x+2}{x^2-4} = \frac{\overset{1}{\cancel{x+2}}}{\underset{1}{\cancel{(x+2)}}(x-2)} = \frac{1}{x-2}$$

Add. Be sure to simplify each answer.

$\dfrac{3}{12} + \dfrac{5}{12} =$	$\dfrac{5}{8} + \dfrac{-1}{8} =$
$\dfrac{5}{9} + \dfrac{6}{9} + \dfrac{7}{9} =$	$\dfrac{4}{3t} + \dfrac{2}{3t} =$
$\dfrac{3}{2x} + \dfrac{7}{2x} =$	$\dfrac{-5}{4x} + \dfrac{7}{4x} =$
$\dfrac{3y}{y+4} + \dfrac{12}{y+4} =$	$\dfrac{3x}{4x+12} + \dfrac{9}{4x+12} =$

$$\frac{a}{a^2-1} + \frac{1}{a^2-1} =$$

$$\frac{x}{x^2+6x+8} + \frac{4}{x^2+6x+8} =$$

$$\frac{5n}{n^2+2n} + \frac{5n}{n^2+2n} =$$

$$\frac{b^2}{b^2+2b-3} + \frac{-9}{b^2+2b-3} =$$

Add. Make sure each answer is in simplest form.

$$\frac{3}{y} + \frac{-8}{y} + \frac{2}{y} =$$

$$\frac{3x}{2x} + \frac{5y}{2x} =$$

$$\frac{x^2}{x^2+6x} + \frac{5x}{x^2+6x} + \frac{-6}{x^2+6x} =$$

$$\frac{x^2}{x^2-9} + \frac{3x}{x^2-9} =$$

$$\frac{-5}{x+4} + \frac{-3}{x+4} =$$

$$\frac{3x^2}{x+4} + \frac{12x}{x+4} =$$

$$\frac{5x^2}{x^2-4x+3} + \frac{-5x}{x^2-4x+3} =$$

$$\frac{11x}{3x-6} + \frac{4x}{3x-6} =$$

$$\frac{-x}{x^2+4} + \frac{3x}{x^2+4} + \frac{-2x}{x^2+4} =$$

$$\frac{2a}{a^2+2a+1} + \frac{2}{a^2+2a+1} =$$

The numerators of these rational expressions are polynomials. Add. Then simplify the answer if you can.

$$\frac{x^2 + 5}{x - 3} + \frac{2x^2 - 4}{x - 3} = \frac{3x^2 + 1}{x - 3}$$

add   add

This can't be factored, so the answer can't be simplified.

$$\frac{x + 10}{x + 2} + \frac{x - 6}{x + 2} =$$

$$\frac{2x + 3}{x + 1} + \frac{x + 4}{x + 1} =$$

$$\frac{a + 2b}{a + b} + \frac{2a + b}{a + b} =$$

$$\frac{y^2 + 4}{y + 1} + \frac{y^2 + 2y}{y + 1} =$$

$$\frac{2 - y^2}{3y} + \frac{y^2 + 7}{3y} =$$

$$\frac{2x - 3}{x + 6} + \frac{x - 1}{x + 6} =$$

$$\frac{x^2 + 7x}{3x(x + 2)} + \frac{2x^2 - x}{3x(x + 2)} =$$

$$\frac{x^2 + 4x + 3}{(x + 1)(x - 3)} + \frac{-x^2 - x}{(x + 1)(x - 3)} =$$

$$\frac{2x^2 + 5}{(x + 1)(x - 3)} + \frac{7 + x - x^2}{(x + 1)(x - 3)} =$$

The Opposite of a Fraction

Subtracting rational numbers and expressions works the same way as subtracting integers. To subtract, we add the opposite — so we need to know how to find the opposite of a fraction.

The **opposite of a fraction** has the *opposite numerator* but the *same denominator* as the original fraction.

$$-\frac{3}{4} = \frac{^-3}{4}$$ "The opposite of $\frac{3}{4}$ is $\frac{^-3}{4}$."

$$-\frac{^-6}{x} = \frac{6}{x}$$ "The opposite of $\frac{^-6}{x}$ is $\frac{6}{x}$."

$$-\frac{x-1}{x^2+3} = \frac{^-x+1}{x^2+3}$$ "The opposite of $\frac{x-1}{x^2+3}$ is $\frac{^-x+1}{x^2+3}$."

Write the opposite of each fraction.

$$-\frac{5}{8} =$$ $$-\frac{x-1}{3} =$$ $$-\frac{a+3}{a-3} =$$

$$-\frac{^-5}{8} =$$ $$-\frac{x+1}{3} =$$ $$-\frac{x^2-5}{x-1} =$$

$$-\frac{3}{x} =$$ $$-\frac{2x+4}{x} =$$ $$-\frac{x^2+2x-3}{x} =$$

$$-\frac{^-3}{x} =$$ $$-\frac{2x-4}{x} =$$ $$-\frac{3x^2-x+4}{x^2-1} =$$

$$-\frac{^-x}{6} =$$ $$-\frac{4}{x-2} =$$ $$-\frac{5}{x^2-2x+1} =$$

Add.

$$\frac{5}{8} + \frac{^-5}{8} =$$ $$\frac{2x+7}{5} + \frac{^-2x-7}{5} =$$

$$\frac{^-x}{2} + \frac{x}{2} =$$ $$\frac{x^2-4x+3}{x+1} + \frac{^-x^2+4x-3}{x+1} =$$

6

Subtracting Fractions with a Common Denominator

To subtract one rational expression from another:

1. Replace the subtraction sign with an addition sign.
2. Replace the *second* fraction with its opposite.
3. Go ahead and add.

Subtract.

$\dfrac{2x}{5} + \dfrac{-3x}{5} = \dfrac{-x}{5}$	$\dfrac{6}{8} - \dfrac{3}{8} =$	$\dfrac{12}{13x} - \dfrac{5}{13x} =$
$\dfrac{2}{15} - \dfrac{11}{15} =$	$\dfrac{x}{4} - \dfrac{5}{4} =$	$\dfrac{2x}{9} - \dfrac{4x}{9} =$
$\dfrac{8}{x+5} - \dfrac{5}{x+5} =$	$\dfrac{9}{4x} - \dfrac{3x}{4x} =$	$\dfrac{x}{x+y} - \dfrac{y}{x+y} =$
$\dfrac{x+5}{6} - \dfrac{x}{6} =$	$\dfrac{x+3}{9} - \dfrac{4}{9} =$	$\dfrac{x+4}{x} - \dfrac{4}{x} =$

$\dfrac{3x+8}{2} + \dfrac{-2x+5}{2} = \dfrac{x+3}{2}$	$\dfrac{2x+1}{x} - \dfrac{x-2}{x} =$
$\dfrac{6x}{x+1} - \dfrac{2x+3}{x+1} =$	$\dfrac{3r+1}{2r} - \dfrac{3r-1}{2r} =$
$\dfrac{a+4}{a-2} - \dfrac{2a-6}{a-2} =$	$\dfrac{x^2-5}{x+3} - \dfrac{x-5}{x+3} =$

$\dfrac{x^2+3x-1}{2x} - \dfrac{x^2-5x+2}{2x} =$

Subtract. Simplify the answer if you can.

$$\frac{3x-4}{3x+3} - \frac{2x-5}{3x+3} =$$

$$\frac{x+y}{2x} - \frac{x-y}{2x} =$$

$$\frac{3x+2}{x} - \frac{4x-5}{x} =$$

$$\frac{3x}{4x-12} - \frac{9}{4x-12} =$$

$$\frac{7x}{x^2+5x} - \frac{3x}{x^2+5x} =$$

$$\frac{x^2+4x}{x+4} - \frac{4x-5}{x+4} =$$

$$\frac{a+5}{ab} - \frac{a-5}{ab} =$$

$$\frac{-x}{15} - \frac{4x}{15} =$$

$$\frac{6a}{5a^2+a} - \frac{a-1}{5a^2+a} =$$

$$\frac{x^2+3x-5}{10} - \frac{x^2-2x+10}{10} =$$

Here are some more rational expressions for you to add. This time you will have to multiply the factors in each numerator to find out if there are any like terms you can combine.

$$\frac{5(x-3)}{7} + \frac{3(x+6)}{7} = \frac{5x-15}{7} + \frac{3x+18}{7} = \frac{8x+3}{7}$$

$$\frac{x(x-3)}{4} + \frac{3(x-2)}{4} =$$

$$\frac{7(x+2)}{12x} + \frac{5(x+2)}{12x} =$$

$$\frac{5(x+4)}{10x} + \frac{2(x-3)}{10x} =$$

$$\frac{b(a+1)}{ab} + \frac{a(2b-3)}{ab} =$$

$$\frac{3(5a)}{2a+2} + \frac{3(a+6)}{2a+2} =$$

$$\frac{x(x-1)}{x(x-3)} + \frac{6(x-4)}{x(x-3)} =$$

$$\frac{3(x+4)}{x^2-16} + \frac{3(x-4)}{x^2-16} =$$

$$\frac{6(a+1)}{a^2-4} + \frac{a(a-1)}{a^2-4} =$$

$$\frac{(x+7)(x-2)}{(x+4)(x-3)} + \frac{(x-5)(x+2)}{(x+4)(x-3)} =$$

These subtraction problems will be easiest if you multiply the expressions in the numerators first. Then change each problem to an equivalent addition problem.

$$\frac{3(x+1)}{5} - \frac{2(x-3)}{5} = \frac{3x+3}{5} - \frac{2x-6}{5} = \frac{3x+3}{5} + \frac{-2x+6}{5} = \frac{x+9}{5}$$

$$\frac{5x}{8} - \frac{4(x+1)}{8} =$$

$$\frac{5(x-4)}{2x} - \frac{4(x-5)}{2x} =$$

$$\frac{2(x+4)}{x} - \frac{6(x-1)}{x} =$$

$$\frac{x^2}{12} - \frac{(x+1)(x-1)}{12} =$$

$$\frac{5(x-1)}{x+2} - \frac{3(x-3)}{x+2} =$$

$$\frac{x(x+1)}{6} - \frac{(x+3)(x-3)}{6} =$$

$$\frac{16}{x} - \frac{2(x+4)}{x} =$$

$$\frac{2a+b}{ab} - \frac{2(a-b)}{ab} =$$

$$\frac{2c(c+3)}{c-5} - \frac{c(c+4)}{c-5} =$$

$$\frac{x^2+1}{y^2} - \frac{(x+1)(x-1)}{y^2} =$$

Adding and Subtracting Fractions with Different Denominators

Now we will add and subtract some fractions with different denominators.
To do this, we first have to find equivalent fractions with a common denominator.

In each problem below you first need to rewrite one of the fractions so that the two fractions have a common denominator. Then you can go ahead and combine.

$\dfrac{2\cdot 2x}{2\cdot 3} + \dfrac{x}{6} = \dfrac{4x}{6} + \dfrac{x}{6} = \dfrac{5x}{6}$	$\dfrac{1}{4} + \dfrac{3}{8} =$
$\dfrac{2}{3} - \dfrac{3}{21} =$	$\dfrac{4}{15} + \dfrac{2}{5} =$
$\dfrac{x}{8} + \dfrac{5x}{16} =$	$\dfrac{4}{5x} - \dfrac{4}{15x} =$
$\dfrac{x}{3x} + \dfrac{2}{x} =$	$\dfrac{5}{2x} + \dfrac{9}{x} =$

$\dfrac{2}{ab} + \dfrac{3}{a^2 b^2} =$

$\dfrac{5x+6}{x^2} - \dfrac{5}{x} =$

$\dfrac{2x}{3} + \dfrac{x}{9} + \dfrac{5x}{6} =$

$\dfrac{2}{a} + \dfrac{3}{ab} + \dfrac{4}{b} =$

$\dfrac{1}{y} - \dfrac{4}{y^2} + \dfrac{3}{y} =$

Add or subtract.

$$\frac{3}{x+2} + \frac{x}{2(x+2)} = \frac{2 \cdot 3}{2(x+2)} + \frac{x}{2(x+2)} = \frac{6}{2(x+2)} + \frac{x}{2(x+2)} = \frac{6+x}{2(x+2)}$$

$$\frac{5}{x-7} + \frac{3}{4(x-7)} =$$

$$\frac{5x}{(x+4)(x-4)} + \frac{1}{x-4} =$$

$$\frac{3}{x+5} + \frac{6}{(x+2)(x+5)} =$$

$$\frac{6}{7} - \frac{5}{7x-14} = \frac{6(x-2)}{7(x-2)} - \frac{5}{7(x-2)} = \frac{6x-12}{7(x-2)} + \frac{-5}{7(x-2)} = \frac{6x-17}{7(x-2)}$$
$$\frac{}{7(x-2)}$$

$$\frac{2}{x} + \frac{5}{x^2 + 2x} =$$

$$\frac{3x+1}{x^2 + 6x + 9} - \frac{2}{x+3} =$$

$$\frac{10}{x-6} - \frac{4}{x^2 - 36} =$$

$$\frac{x}{x^2 - 9} + \frac{2}{x-3} =$$

$$\frac{10a}{a^2 + 6a} - \frac{3}{a+6} =$$

$$\frac{3x}{2x^2 - 8x} + \frac{2}{x-4} =$$

Sometimes denominators that look different are really equivalent or opposites. Be on the lookout for these. If they are opposites, it is easy to make both denominators the same. All we have to do is to multiply one fraction by $\frac{-1}{-1}$.

Add or subtract. Look for denominators that are equivalent or opposites.

$$\frac{3}{2x+5} + \frac{4x}{5+2x} = \frac{3+4x}{5+2x}$$

$$\underbrace{}_{\text{equivalent}}$$

$$\frac{7x}{x-3} + \frac{5x}{3-x} = \frac{7x}{x-3} + \frac{5x(-1)}{(3-x)(-1)} = \frac{7x}{x-3} + \frac{-5x}{x-3} = \frac{2x}{x-3}$$

$$\underbrace{}_{\text{opposites}}$$

$$\frac{4}{3x+1} + \frac{3}{1+3x} =$$

$$\frac{5}{2-x} + \frac{x}{x-2} =$$

$$\frac{2x}{3x-1} + \frac{5x}{1-3x} =$$

$$\frac{7x}{x^2-1} + \frac{7}{1-x^2} =$$

$$\frac{2x+8}{x+5} - \frac{x+3}{5+x} =$$

$$\frac{x}{x-3} + \frac{2x}{-x+3} =$$

$$\frac{12}{4-x} - \frac{x+8}{x-4} =$$

Combining Integers and Rational Expressions

In arithmetic we can write $2 + \frac{3}{4}$ as $2\frac{3}{4}$ or we can combine these into a single fraction, $\frac{11}{4}$. In algebra we can also combine an integer with a rational expression.

Combine.

$1 + \frac{3}{x} = \frac{1 \cdot x}{x} + \frac{3}{x} = \frac{x}{x} + \frac{3}{x} = \frac{x+3}{x}$	$1 + \frac{x}{3} =$
$2 - \frac{3}{4} =$	$8 - \frac{3}{4} =$

$2 + \frac{3}{x-1} = \frac{2(x-1)}{x-1} + \frac{3}{x-1} = \frac{2x-2}{x-1} + \frac{3}{x-1} = \frac{2x+1}{x-1}$

$5 + \frac{1}{x+2} =$

$3 + \frac{x}{x-1} =$

$\frac{2x}{x+3} - 1 =$

$\frac{x+4}{x+2} + 2 =$

$\frac{x}{y} + 4 =$	$\frac{x}{x-y} + 1 =$
$1 + \frac{1}{x} - \frac{1}{x^2} =$	$1 + \frac{1}{x} + \frac{1}{x^2} =$
$\frac{1}{x^2} - \frac{2}{x} + 3 =$	$\frac{1}{4} - 1 + \frac{3}{4} =$

14

Combining Polynomials and Rational Expressions

We can combine polynomials and rational expressions, too. First we have to write the polynomial as a rational expression with the same denominator as the other expression.

Combine.

$$5y + \frac{2}{y} = \frac{y \cdot 5y}{y} + \frac{2}{y} = \frac{5y^2}{y} + \frac{2}{y} = \frac{5y^2 + 2}{y}$$

$$2x + \frac{1}{x} =$$

$$a + \frac{a}{b} =$$

$$b + \frac{a}{b} =$$

$$(x + 1) + \frac{3}{x} =$$

$$\frac{x}{3} + (x - 5) =$$

$$(x + 3) + \frac{2x}{x-1} = \frac{(x+3)(x-1)}{x-1} + \frac{2x}{x-1} = \frac{x^2 + 2x - 3}{x-1} + \frac{2x}{x-1} = \frac{x^2 + 4x - 3}{x-1}$$

$$(x + 1) - \frac{x^2}{x+1} =$$

$$(3x - 4) + \frac{x^2}{2} =$$

$$(2x - 3) - \frac{1}{2x-3} =$$

More Adding and Subtracting Fractions with Different Denominators

For each problem on this page you first need to rewrite *all* the fractions so that they have a common denominator. You can always find a common denominator by multiplying the denominators together.

$\dfrac{3 \cdot 2}{3 \cdot 5} + \dfrac{1 \cdot 5}{3 \cdot 5} = \overline{15} + \overline{15} =$	$\dfrac{1}{2} + \dfrac{2}{3} =$
$\dfrac{4}{7} - \dfrac{1}{8} =$	$\dfrac{3}{8} - \dfrac{7}{9} =$
$\dfrac{2x}{5} - \dfrac{x}{4} =$	$\dfrac{7}{y} + \dfrac{5}{x} =$
$\dfrac{1}{3} + \dfrac{5}{14} =$	$\dfrac{7}{a} + \dfrac{3}{b} =$
$\dfrac{4}{x} - \dfrac{3}{2y} =$	$\dfrac{x}{3} - \dfrac{4}{5} =$
$\dfrac{3x}{5} + \dfrac{4}{x} =$	$\dfrac{8}{x} - \dfrac{7}{y} =$
$\dfrac{x}{3} + \dfrac{x}{2} + \dfrac{x}{5} =$	$\dfrac{3}{a} + \dfrac{4}{b} + \dfrac{1}{c} =$
$\dfrac{9}{4x} + \dfrac{x}{3} =$	$\dfrac{2}{x^2} - \dfrac{1}{y^2} =$
$\dfrac{3x}{5y} + \dfrac{5y}{3x} =$	$\dfrac{4}{a} - \dfrac{2}{3} =$
$\dfrac{5}{12} - \dfrac{1}{x^2} =$	$\dfrac{8}{3} + \dfrac{7}{2x} =$

Find a common denominator. Then add or subtract.

$$\frac{3x}{7} + \frac{2}{x+5} = \frac{3x(x+5)}{7(x+5)} + \frac{7\cdot 2}{7(x+5)} = \frac{3x^2+15x}{7(x+5)} + \frac{14}{7(x+5)} = \frac{3x^2+15x+14}{7(x+5)}$$

$$\frac{3}{8} - \frac{5}{a-2} =$$

$$\frac{5}{x+2} + \frac{3}{4x} =$$

$$\frac{2}{x+1} + \frac{2}{5x} =$$

$$\frac{x-1}{x+3} + \frac{2}{x+1} =$$

$$\frac{x+4}{x} - \frac{x}{x+3} =$$

$$\frac{x+3}{x+5} + \frac{x}{x+2} = \frac{(x+2)(x+3)}{(x+2)(x+5)} + \frac{x(x+5)}{(x+2)(x+5)} = \frac{x^2+5x+6}{(x+2)(x+5)} + \frac{x^2+5x}{(x+2)(x+5)} = \frac{2x^2+10x+6}{(x+2)(x+5)}$$

$$\frac{x+4}{x-4} + \frac{3}{x+3} =$$

$$\frac{x-4}{x+6} - \frac{x}{x+7} =$$

$$\frac{2x}{2x+1} + \frac{x-1}{x+4} =$$

$$\frac{x-5}{x+1} - \frac{x+3}{x-1} =$$

Least Common Denominators

The product of the two denominators is not always the least or simplest denominator we could use. Think about adding $\frac{5}{12}$ and $\frac{7}{8}$.

If we use 12 times 8, or 96, as a common denominator, we get

$$\frac{5}{12} + \frac{7}{8} = \frac{5 \cdot 8}{12 \cdot 8} + \frac{12 \cdot 7}{12 \cdot 8} = \frac{40}{96} + \frac{84}{96} = \frac{124}{96}$$

We could have used 24 as a common denominator.

$$\frac{5}{12} + \frac{7}{8} = \frac{5 \cdot 2}{12 \cdot 2} + \frac{3 \cdot 7}{3 \cdot 8} = \frac{10}{24} + \frac{21}{24} = \frac{31}{24}$$

The answers are equivalent, but $\frac{31}{24}$ is simpler.

Add or subtract each pair of fractions. Try to find the **least common denominator**.

$$\frac{11}{6} - \frac{4}{15} =$$

$$\frac{3}{20} + \frac{8}{45} =$$

$$\frac{a}{20} - \frac{b}{24} =$$

$$\frac{1}{12x} + \frac{5}{8x} =$$

$$\frac{3}{2x} + \frac{2}{x^2} =$$

$$\frac{5}{2x^2} + \frac{3}{6xy} =$$

$$\frac{1}{x^2} + \frac{4}{xy} =$$

$$\frac{5}{6x} - \frac{1}{10} =$$

Did you discover a way to find the least common denominator? Here is a method that always works. We factor the denominators. Then we multiply each denominator by the factors of the other denominator which are missing. We multiply the numerator by those factors, too.

Find the least common denominator. Then combine.

$$\frac{5}{12} + \frac{7}{8} = \frac{5}{2 \cdot 2 \cdot 3} + \frac{7}{2 \cdot 2 \cdot 2} = \frac{5 \cdot 2}{2 \cdot 2 \cdot 3 \cdot 2} + \frac{7 \cdot 3}{2 \cdot 2 \cdot 2 \cdot 3} = \frac{10}{24} + \frac{21}{24} = \frac{31}{24}$$

The other denominator has 3 2's. This one needs another **2**.

The other denominator has a 3, so this one needs a **3**.

$$\frac{8}{35} + \frac{2}{45} =$$

$$\frac{2}{9} - \frac{2}{15} =$$

$$\frac{3}{ab} + \frac{4}{b^2c} = \frac{3}{ab} + \frac{4}{bbc} = \frac{3 \cdot bc}{ab \cdot bc} + \frac{4 \cdot a}{a \cdot bbc} = \frac{3bc}{ab^2c} + \frac{4a}{ab^2c} = \frac{3bc + 4a}{ab^2c}$$

Needs a **b** and a **c**.

Needs an **a**.

$$\frac{2}{9x^2} + \frac{4}{15x} =$$

$$\frac{8}{abc} - \frac{2}{bcd} =$$

$$\frac{1}{14x^2} + \frac{1}{21xy} =$$

$$\frac{3}{x} + \frac{2}{x^2} - \frac{1}{x^3} =$$

$$\frac{5}{a^2b} - \frac{3}{ab^2} =$$

Find the least common denominator. Then add or subtract.

$$\frac{x}{4x+6} - \frac{2}{6x+9} = \frac{3 \cdot x}{3 \cdot 2(2x+3)} - \frac{2 \cdot 2}{2 \cdot 3(2x+3)} = \frac{3x}{6(2x+3)} + \frac{-4}{6(2x+3)} = \frac{3x-4}{6(2x+3)}$$

$2(2x+3)\quad 3(2x+3)$

$$\frac{1}{x^2-3x} + \frac{5}{4x-12} =$$

$$\frac{3}{x-1} + \frac{4}{x^2-1} =$$

$$\frac{1}{3x-6} - \frac{x}{x^2-4} =$$

$$\frac{3}{x^2-25} + \frac{5}{2x+10} =$$

$$\frac{5}{x^2-9} + \frac{2}{x^2-6x+9} =$$

$$\frac{4}{n^2-4} + \frac{2}{n^2-5n+6} =$$

$$\frac{3}{x^2+6x+5} + \frac{1}{x^2+4x+3} =$$

$$\frac{5}{3x^2-12} + \frac{2}{4x+8} =$$

Add.

$\dfrac{4}{x-6} + \dfrac{3}{6-x} =$

$\dfrac{7}{a+2} + \dfrac{4}{a+5} =$

$\dfrac{x}{x+3} + \dfrac{2}{5x+15} =$

$\dfrac{10}{x^2y} + \dfrac{4}{xy^2} =$

$\dfrac{1}{x^2-3x} + \dfrac{3}{2x-6} =$

$\dfrac{x}{y} + 4 =$

$\dfrac{3}{a} + \dfrac{2}{b} + \dfrac{4}{ab} =$

$\dfrac{x}{x^2-1} + \dfrac{2}{x^2+6x-7} =$

$\dfrac{y}{3y+1} + \dfrac{2}{1+3y} =$

Find a common denominator and add. Then simplify.

$$\frac{x-2}{4} + \frac{x+2}{12} =$$

$$\frac{2x+1}{20} + \frac{x+3}{15} =$$

$$\frac{x+6}{x+5} + \frac{x+3}{2x+10} =$$

$$\frac{3}{x-2} + \frac{x-11}{3x-6} =$$

$$\frac{a+b}{ab} + \frac{a-2b}{2ab} =$$

$$\frac{x+2}{x-1} + \frac{x-4}{x^2-x} =$$

$$\frac{x}{2x-5} + \frac{5-x}{5-2x} =$$

$$\frac{8}{4x+4} + \frac{x+3}{x^2+x} =$$

$$\frac{x-6}{x^2-5x} + \frac{1}{5x-25} =$$

Using Common Denominators to Solve Equations

Common denominators are useful for solving equations which contain rational expressions. In the equation below there are three fractions. The least common denominator for these fractions is $6x^2$. If we multiply both sides of the equation by the common denominator, we will get an equivalent equation without fractions.

$$\frac{1}{2x} + \frac{2}{x^2} = \frac{7}{6x}$$

The least common denominator is $6x^2$.

$$6x^2\left(\frac{1}{2x} + \frac{2}{x^2}\right) = \frac{7}{6x} \cdot 6x^2$$

$$\frac{\overset{3x}{6x^2}}{1} \cdot \frac{1}{2x} + \frac{6x^2}{1} \cdot \frac{2}{x^2} = \frac{7}{6x} \cdot \frac{\overset{x}{6x^2}}{1}$$

$$3x + 12 = 7x$$
$$12 = 4x$$
$$x = 3$$

3 works: $\frac{1}{6} + \frac{2}{9} = \frac{7}{18}$ is true, so it's a solution.

Solve each equation by multiplying both sides by the least common denominator.

$$\frac{1}{3} + \frac{2}{x} = \frac{2}{3}$$

$$\frac{x}{5} + \frac{x}{2} = \frac{14}{5}$$

$$\frac{x}{3} + \frac{2x}{3} = 4$$

$$\frac{3a}{4} + \frac{a}{6} = 0$$

Solve each equation.

$$\frac{x+1}{8} = \frac{x}{2} - \frac{1}{4}$$

$$\frac{x+2}{8} - \frac{x}{12} = \frac{1}{2}$$

$$\frac{3}{2x} - \frac{1}{x} = \frac{1}{4}$$

$$x + \frac{24}{x} = 11$$

$$5 + \frac{3}{x} = \frac{7}{2}$$

$$\frac{x}{2} - \frac{7}{2} = \frac{9}{x}$$

$$\frac{3}{x-1} = \frac{1}{2} - \frac{4}{x-1}$$

$$\frac{x}{3} + \frac{10}{3(x+4)} = \frac{5}{x+4}$$

Here are some more equations with rational expressions for you to solve.

$$12\left(\frac{3}{4}x + \frac{1}{3}x\right) = (x+1)12$$

$$\frac{2}{5}x - \frac{1}{10}x = 1$$

$$0.8x = 2 + 0.7x$$

This says $\frac{8}{10}x = 2 + \frac{7}{10}x$, so multiply both sides by 10.

$$10(0.8x) = (2 + 0.7x)10$$
$$8x = 20 + 7x$$
$$x = 20$$

$$0.5 + 0.1x = 0.8x - 0.9$$

$$0.23x + 2.25 = 0.45x - 1.05$$

$$0.7x - 0.66x = 2$$

$$0.2x - 0.08x = 1.2$$

$$1.24x = 3.2 + 0.6x$$

Proportions

A **proportion** is a very simple rational equation which says that one fraction equals another. Each of these is a proportion:

$$\frac{3}{4} = \frac{15}{20} \qquad\qquad \frac{x}{2} = \frac{5}{9} \qquad\qquad \frac{x+1}{x} = \frac{4}{x+2}$$

Look at what happens to a proportion if we get a common denominator by multiplying the two denominators together and then use it to simplify the equation.

$$\frac{3}{4} = \frac{15}{20} \qquad\qquad\qquad \frac{x}{2} = \frac{5}{9}$$

$$\overset{20}{\cancel{80}} \cdot \frac{3}{4} = \frac{15}{\cancel{20}} \cdot \overset{4}{\cancel{80}} \qquad\qquad \overset{9}{\cancel{18}} \cdot \frac{x}{2} = \frac{5}{\cancel{9}} \cdot \overset{2}{\cancel{18}}$$

$$20 \cdot 3 = 15 \cdot 4 \qquad\qquad\qquad 9x = 5 \cdot 2$$

It looks as if each denominator has "moved" to the other side of the equation.

$$\frac{3}{4} \diagup\!\!\!\!\diagdown \frac{15}{20} \qquad\qquad\qquad \frac{x}{2} \diagup\!\!\!\!\diagdown \frac{5}{9}$$

$$20 \cdot 3 = 15 \cdot 4 \qquad\qquad\qquad 9x = 5 \cdot 2$$

The arrows show where each denominator ended up. Since this is always the pattern, we usually skip the first step and go right to multiplying each numerator by the opposite denominator. We call this **cross multiplying**.

Cross multiplying makes it easy to solve proportions. Use cross multiplying to solve each proportion below.

$\frac{12}{16} = \frac{x}{20}$ $20 \cdot 12 = 16 \cdot x$ $240 = 16x$ $x = 15$	$\frac{-10}{8} = \frac{x}{12}$	$\frac{20}{8} = \frac{15}{x}$
$\frac{x}{4} = \frac{3}{8}$	$\frac{4}{x} = \frac{7}{9}$	$\frac{8}{3} = \frac{x+1}{6}$

Solve each proportion.

$\dfrac{10}{x} = \dfrac{15}{x-6}$ $10(x-6) = 15 \cdot x$ $10x - 60 = 15x$ $-60 = 5x$ $x = -12$	$\dfrac{x}{6} = \dfrac{x+8}{18}$	$\dfrac{8}{x} = \dfrac{14}{x+3}$
$\dfrac{3}{7a} = \dfrac{2}{5}$	$\dfrac{p}{p+4} = \dfrac{7}{8}$	$\dfrac{3}{n+2} = \dfrac{-3}{n-2}$
$\dfrac{4}{5} = \dfrac{y-3}{y+3}$	$\dfrac{x+3}{x-1} = \dfrac{x-5}{x-7}$	$\dfrac{x}{x+6} = \dfrac{x-2}{x+2}$
$\dfrac{x}{4} = \dfrac{1}{x}$ $x^2 = 4$ $x^2 - 4 = 0$ $(x-2)(x+2) = 0$ $x - 2 = 0 \mid x + 2 = 0$ $x = 2 \mid x = -2$	$\dfrac{x+1}{3} = \dfrac{2}{x}$	$\dfrac{x}{4} = \dfrac{9}{x}$
$\dfrac{x+6}{-3} = \dfrac{4}{x-1}$	$\dfrac{x+7}{3} = \dfrac{5}{x-7}$	$\dfrac{20}{x} = \dfrac{x}{5}$

Ratio Problems

We can use proportions to solve many problems that involve ratios. A **ratio** is a fraction which compares one number to another. The ratio of 2 to 3 is $\frac{2}{3}$. We can write a proportion if we know that two ratios are equivalent.

Write a proportion for each problem. Then solve the proportion to find the answer.

If it takes 5 gallons of gas to drive 90 miles, how many gallons would it take to drive 144 miles?

The ratio of gallons to miles is always the same.

Equation:
$$\frac{5}{90} = \frac{x}{144}$$
$$144 \cdot 5 = 90 \cdot x$$
$$720 = 90x$$
$$x = 8$$

Answer: **It takes 8 gallons to drive 144 miles.**

The cafeteria used 28 bottles of ketchup in 18 days. How much ketchup should be ordered for 45 days?

Equation:

Answer:

A ball player gets 11 hits in 40 times at bat. How many hits would you predict in 1000 times at bat?

Equation:

Answer:

The ratio of sand to cement in concrete is 5 to 2. How many shovels of sand should be mixed with 8 shovels of cement?

Equation:

Answer:

A typist can type 300 words in 4 minutes. At that rate, how long would it take to type a 13,500 word document?

Equation:

Answer:

If 12 lb. of coleslaw will feed 100 people, how much would you need to feed 40 people?

Equation:

Answer:

Solve each problem by using a proportion.

The ratio of red to brown candies in a well-known brand is 2 to 3. Chris likes the red ones. Jamie prefers brown. How many will Chris be likely to get if Jamie gets 45?

Equation:

Answer:

The label on a 48-lb. bag of lawn fertilizer says it covers 15,000 square feet. How many pounds will be needed to cover 50,000 square feet?

Equation:

Answer:

A car's trip meter registers 10.2 miles for a 10-mile measured course. When the meter registers 153 miles, what distance has the car actually traveled?

Equation:

Answer:

A cake recipe calls for 2 eggs to $1\frac{1}{2}$ cups of flour. How many eggs should be used with 9 cups of flour?

Equation:

Answer:

The ratio of inches to centimeters in a measurement is about 2 to 5. Maria's waist measures 70 centimeters. About what would that be in inches?

Equation:

Answer:

The ratio of left-handed people to all people is about 1 to 10. How many left-handed students would you expect to find in a high school with 450 students?

Equation:

Answer:

Percent Problems

The word **percent** means "hundredths" or "out of a hundred." A percent is a ratio with a denominator of 100 and so can be written as either a fraction or a decimal.

73%, $\frac{73}{100}$ and .73 all stand for the ratio of 73 to 100.

Often the easiest way to solve a percent problem is to write a proportion.

Write a proportion for each problem. Then solve the proportion to get the answer.

---

32 out of 80 people in line got tickets for the concert. What percent of the people in line were successful?

32 out of 80 is equivalent to what number out of 100?

Equation: $\dfrac{32}{80} = \dfrac{x}{100}$

$100 \cdot 32 = 80 \cdot x$

$3200 = 80x$

$x = 40$

Answer: **40% of the people were successful in getting tickets.**

Concert tickets cost $26 each. 22% went to charity. How much went to charity?

Equation:

Answer:

$15 of Dan's $125 wages was withheld for taxes. What percent of his wages was withheld?

Equation:

Answer:

If 12% is always withheld from Dan's wages, how much would be taken out if he earned $220?

Equation:

Answer:

A survey has shown that 20% of teenagers think they are overweight. In a class of 325, how many would this be?

Equation:

Answer:

The price of gasoline was 96¢ a gallon before it went up 25%. How much did it go up?

Equation:

Answer:

Solve each problem using a proportion.

A jacket is on sale for $68, which is 80% of its original price. What was the price before the sale? *Equation:* *Answer:*	Thea would like to leave a 15% tip. Her bill was $5.40. How much should she leave? *Equation:* *Answer:*
The number of students at Plainview High went from 520 to 624 in just one year. This year's student body is what percent of last year's? *Equation:* *Answer:*	Ted read on a cereal box that a serving contains 3 grams of protein, which is 4% of the protein a person should eat each day. According to this, how many grams of protein should a person eat daily? *Equation:* *Answer:*
Read the problem above again. This time figure out what percent last year's student body was of this year's. *Equation:* *Answer:*	In the science lab Shad found that a candle used 60 out of 270 ml of trapped air before going out. What percent of the air did the candle use? *Equation:* *Answer:*

Time Problems

How long *would* it take Jack and Jan working together? We can write an equation to find the answer. All we have to do is to think about how much of the work each one can do in an hour.

	Total Time	*Amount Done in an Hour*
Jack	6 hours	$\frac{1}{6}$
Jan	5 hours	$\frac{1}{5}$
together	x hours	$\frac{1}{x}$

Think: If we add the amounts each one does in an hour, we should get the amount they do together in an hour.

Equation:

$$\frac{1}{6} + \frac{1}{5} = \frac{1}{x}$$

$$30x\left(\frac{1}{6} + \frac{1}{5}\right) = \left(\frac{1}{x}\right)30x$$

$$5x + 6x = 30$$

$$11x = 30$$

$$x = \frac{30}{11} = 2\frac{8}{11}$$

Answer: Jan was right! They can do the job together in less than three hours.

Write an equation for each problem. Then solve it to find the answer.

Jon delivers all his newspapers in 3 hours. It takes his trainee 4 hours to cover the route. How long will it take them together if they start from opposite ends?	Rex eats a bag of dog food in 12 days. Duke goes through a bag in 6 days. How long will one bag feed the two of them?

Equation:

Equation:

Answer:

Answer:

Tracy can mow the lawn with a power mower in 20 minutes. Hai can mow it with a hand mower in 50 minutes. How long will it take them if they cooperate?	The old duplicator could produce an issue of the newsletter in 2 hours. The new one takes 40 minutes. How long will it take if both machines are used?

Equation:

Equation:

Answer:

Answer:

Problems about Rational Numbers

Make up an equation for each problem. Then solve the equation to get the answer.

"I'm thinking of a number. If you add half of this number and a third of this number, you will get 35. What is my number?"

Equation: $\frac{x}{2} + \frac{x}{3} = 35$

$6\left(\frac{x}{2} + \frac{x}{3}\right) = 35 \cdot 6$

$3x + 2x = 210$

$5x = 210$

Answer: 42 $x = \frac{210}{5} = 42$

"I'm thinking of an integer. When half of this integer is added to a third of the next integer, the result is 7. What is the integer?"

Equation:

Answer:

"I'm thinking of a number. If I add this number to its reciprocal, I get 2. What is the number?"

Equation:

Answer:

"If I take a fourth of a number away from half the number, I get 10. What is the number?"

Equation:

Answer:

Diophantus was a famous mathematician of ancient Greece. A legend says there was a number puzzle on his tomb — a puzzle that you can solve the way you solved the other problems on this page.

"This tomb holds Diophantus. For one sixth of his life he was a boy, for one twelfth, a youth. After one seventh more he married and five years later had a son. The son lived only half as long as his father, and died four years before him." How long did Diophantus live?

Equation:

Answer:

Written Work

Do these problems on some clean paper. Label each page of your work with your name, your class, the date, and the book number. Also number each problem. Keep this written work inside your book, and turn it in with your book when you are finished. Please do a neat job.

1. Find the sum, difference, product and quotient of each pair of fractions.

$\frac{5}{8}$ and $\frac{2}{5}$　　　　　$\frac{a}{a+3}$ and $\frac{3}{a+3}$　　　　　$\frac{a}{5b}$ and $\frac{b}{10a^2}$

$\frac{2}{x}$ and $\frac{4}{y}$　　　　　$\frac{4}{n+2}$ and $\frac{4}{n-2}$　　　　　$\frac{x+9}{x-2}$ and $\frac{x-4}{x+3}$

2. Simplify.

$\frac{2}{3} + \frac{4}{5} - \frac{3}{4}$　　　　　$7 - \frac{2}{x+1}$　　　　　$\frac{x+9}{x^2-25} + \frac{4}{x+5}$

$\frac{5}{4} + \frac{10}{x} + \frac{3}{2x^2}$　　　　　$3x - \frac{1}{x}$　　　　　$\frac{x+1}{2x+10} - \frac{x-1}{3x+15}$

$\frac{1}{cd} + \frac{2}{c^2d} + \frac{3}{cd^2}$　　　　　$\frac{4}{x-5} - \frac{5}{x+4}$　　　　　$\frac{2}{x^2-1} + \frac{1}{(x-1)^2} - \frac{1}{(x+1)^2}$

$\frac{x}{y} + x + y$　　　　　$\frac{x}{x+2} + \frac{3}{x^2+2x}$　　　　　$\frac{x}{x^2+2x-48} + \frac{2}{x^2-4x-12}$

$\frac{1}{6x} - \frac{1}{8x} - \frac{1}{10x}$　　　　　$\frac{3}{x-1} - \frac{2x+4}{x^2-1}$

3. Terry and Sandy had another argument. Sandy said that the answer to this problem is 11. Here is Sandy's work:

$$\frac{5}{x} + \frac{1}{2x} = 2x\left(\frac{5}{x} + \frac{1}{2x}\right)$$
$$= \frac{2x}{1}\cdot\frac{5}{x} + \frac{2x}{1}\cdot\frac{1}{2x}$$
$$= 10 + 1$$
$$= 11$$

Terry disagreed with Sandy's answer. Who was right? Explain why.

4. In a survey of 300 students, 117 said they would like to be like their parents. 183 said they definitely did not want to be like their parents. Write a proportion and then solve it to find the answer to each question below.

a. What percent of the students would like to be like their parents?

b. If these results are typical and you interviewed 500 other students, how many would you expect would not want to be like their parents?

Practice Test

Add. Simplify your answer if you can.

$\dfrac{7}{5} + \dfrac{8}{5} =$

$\dfrac{y}{2x} + \dfrac{3}{2x} =$

$\dfrac{6}{x+5} + \dfrac{5}{x+5} =$

$\dfrac{x-2}{x+2} + \dfrac{x+6}{x+2} =$

$\dfrac{8}{a^2} + \dfrac{1}{a} =$

$\dfrac{3}{p} + \dfrac{4}{q} =$

$x + \dfrac{x}{4} =$

$3 + \dfrac{2}{x-5} =$

$\dfrac{2}{x+10} + \dfrac{3}{5x+50} =$

$\dfrac{12}{x^2-9} + \dfrac{2}{x+3} =$

$\dfrac{x}{x+4} + \dfrac{2}{x-1} =$

Subtract. Simplify your answer if you can.

$\dfrac{14}{9} - \dfrac{2}{9} =$

$\dfrac{2}{a-1} - \dfrac{5}{a-1} =$

$\dfrac{x+4}{7} - \dfrac{x-5}{7} =$

$\dfrac{5}{2n} - \dfrac{2}{n} =$

$\dfrac{3}{2x} - \dfrac{4}{6x^2} =$

$\dfrac{3x}{2x+1} - \dfrac{x}{2x+1} =$

$\dfrac{x}{x-3} - \dfrac{6x}{x^2-9} =$

$\dfrac{x+2}{x-4} - \dfrac{3}{x+1} =$

olve each equation.

$$\frac{x}{2} - \frac{x}{4} = 6$$

$$\frac{x}{5} = \frac{3}{7}$$

$$\frac{1}{x} + \frac{3}{2x} = \frac{1}{2}$$

$$\frac{x-1}{x} = \frac{x}{4}$$

Make up an equation for each problem. Then solve the equation to get the answer.

If two inches on a map represents 25 miles, how many miles would 7 inches represent?

Equation:

Answer:

45 of the 75 seniors have jobs.
What percent of the seniors are working?

Equation:

Answer:

Pat can shelve a cart full of books in the library in 24 minutes and Kevin can do it in 30 minutes. How long will it take them if they work together?

Equation:

Answer:

"I'm thinking of a number. If you add half of this number to a sixth of this number, you will get 12. What is the number?"

Equation:

Answer:

Key to Algebra® workbooks

Also available in the Key to...® series

Key to Fractions®
Key to Decimals®
Key to Percents®
Key to Geometry®
Key to Measurement®
Key to Metric Measurement®
The Key to Tracker®, the online companion for the
Key to Algebra, Fractions, Decimals, and Percents workbooks

Key Curriculum Press
INNOVATORS IN MATHEMATICS EDUCATION

ISBN 978-1-55953-007-1